PLUTONIUM/Contents

THE COVER

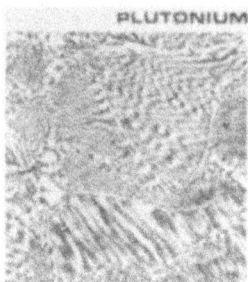

THE AUTHOR

William N. Miner is a staff metallurgist of the Plutonium Physical Metallurgy Group at the Los Alamos Scientific Laboratory, Los Alamos, New Mexico. A graduate of the Colorado School of Mines in 1949, he formerly worked for the Ames Laboratory at the State University of Iowa. Mr. Miner has written numerous technical publications and research reports dealing with plutonium. He was senior author of the section on "Plutonium" in *The Rare Metals Handbook* (1961). He was coauthor of several scientific articles including a technical review on "Physical Metallurgy of Plutonium and Its Alloys" in the *Proceedings of the Second United Nations International Conference on the Peaceful Uses of Atomic Energy, Geneva, 1958.*

PLUTONIUM

By WILLIAM N. MINER

A STRANGE, WONDERFUL SUBSTANCE

Imagine a substance so loaded with energy that a pound of it could give the same explosive effect as 20 million pounds of TNT. Imagine this material also to be a source of power that could be used, even in the present state of knowledge, to supply mankind's energy needs long after the world's resources of conventional fuels such as coal and petroleum have become exhausted.

Build into this imaginary material strange combinations of properties. Under some conditions let it be nearly as hard and brittle as window glass; under others, as soft and plastic as lead. Let it burn and quickly crumble to powder when heated in air, or slowly disintegrate when kept in air at room temperature. Allow its atoms to change continuously to those of a completely different element, so slowly that thousands of years would pass before as much as half of the substance would be changed, but rapidly enough that the heat generated by the reaction would cause the material to feel warm to your hand.

Finally, imagine this substance to be far more poisonous than the cyanide of gas chambers and murder mystery stories, but let people work safely with it day in and day out.

1

A generation ago such a material would have existed only in the imagination of a science fiction writer. Today, we know it as a real element. It is plutonium.

What is plutonium? How was it discovered? How is it formed and produced? What are its properties? Of what use is it, and how do people work with it? This booklet does not attempt to give complete answers to these questions, but it does shed some light on them. For those who wish to delve more deeply into the subject, a list of references is appended. Some of the specialized terms used are explained where they occur, but a glossary has also been included.

THE NATURE OF PLUTONIUM

A Man-made Metal

Plutonium was the first of the man-made elements to be produced in large enough amounts to be visible. (Technetium, discovered in 1937 by C. Perrier and Emilio Segrè, is generally recognized as being the first new element created by man.)

It is a heavy metallic element (more than twice as dense as iron). Its freshly prepared, uncorroded surface has a bright, silvery appearance, which tarnishes rapidly when exposed to air. While most metals are good conductors of electricity and heat, plutonium is not. Its electrical conductivity (ability to conduct electricity) and its thermal conductivity (ability to conduct heat) are both exceptionally low, only about one-hundredth as much as the conductivities of silver.

Its Place Among the Elements

Plutonium is a radioactive element. It has atomic number 94, and its position in the Periodic Table shows that it is the sixth member in the series of elements called the "actinides," of which actinium, atomic number 89, is the first member. Plutonium is also one of the "transuranium" elements, since it has an atomic number higher than that of uranium (number 92). Element 93 and those with higher numbers are man-made by transmutation (changing one element into another). (See Periodic Chart and List of Elements, pages 40 and 41.)

Plutonium's Isotopes

Plutonium has 15 known isotopes;* they range in atomic weight from 232 to 246. All the isotopes are radioactive and decay spontaneously, according to various sequences, eventually forming a stable element such as bismuth or lead. Although all the plutonium isotopes are the same

THE HALF-LIVES OF TWO PLUTONIUM ISOTOPES*

| 0 | 24,000 years | 48,000 years |

Pu-239 — U-235 / Pu-239 — U-235 / Pu-239

1 lb Pu-239 after 24,360 years becomes

½ lb Pu-239 plus ½ lb U-235 after 24,360 years becomes

¼ lb Pu-239 plus ¾ lb U-235 after 24,360 years becomes etc.

| 0 | 10 years | 20 years |

Pu-241 — Am-241 / Pu-241 — Np-237 / Am-241 / Pu-241

1 lb Pu-241 after 10 years becomes

½ lb Pu-241 plus ½ lb Am-241 after 10 years becomes

¼ lb Pu-241 plus nearly ¾ lb Am-241 plus a little Np-237 after 10 years becomes etc.

The Uranium-235 also has a half-life and changes to Thorium-231, but the half-life is so long—over 700 million years—that even after 48,720 years (two half-lives of Plutonium-239) only about one part in 10,000 of the Uranium-235 would have changed to Thorium-231.

Americium-241 has a half-life of 500 years and changes to Neptunium-237. After 20 years (two half-lives of Plutonium-241) a very small but detectable amount of Neptunium-237 would be present with the Plutonium-241 and Americium-241.

*To avoid complicating the illustration, no account was taken of the weight and volume differences between parent and daughter elements. Also, the daughter elements formed in subsequent steps in the radioactive decay process are not shown.

*Isotopes of an element have the same atomic number but different atomic weights.

chemically, the rates at which they disintegrate radioactively differ widely. The half-life of the plutonium-233 isotope, for example, is about 20 minutes while that of the plutonium-244 isotope is about 76 million years. At present, plutonium-239 is the isotope of greatest importance because it is readily fissionable, has the relatively long half-life of 24,360 years, and can be produced in amounts large enough to be of practical use.

Fission

The fissionable property of plutonium makes it an important source of nuclear energy. When a neutron strikes the nucleus of a plutonium-239 atom and sticks to it (rather than glancing off), the nucleus splits (or fissions) to form two atoms of different elements, each having an atomic number about half that of plutonium. (Strontium and xenon or barium and krypton are examples.) At the same time, two or more neutrons are ejected from the plutonium nucleus.

FISSIONING OF PLUTONIUM-239

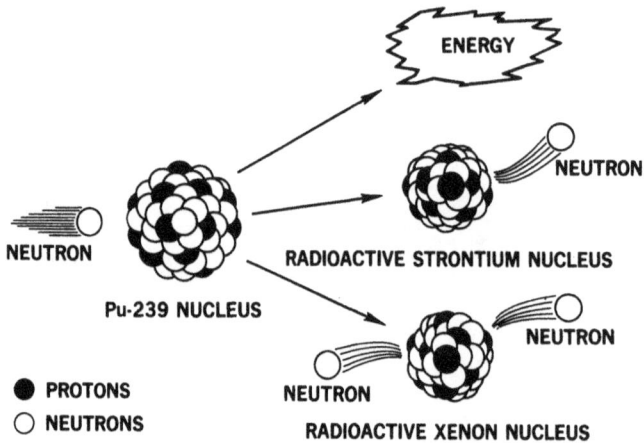

When an atom of plutonium-239 captures a neutron in its nucleus it fissions or divides itself into two radioactive nuclei of very roughly equal size, emits neutrons and liberates a large amount of energy. The energy produced is equivalent to the difference between the masses of (1) a neutron and the plutonium-239 atom and (2) the two newly produced radioactive atoms plus the emitted neutrons.

4

These new neutrons may be absorbed by nuclei of other plutonium atoms, which then fission, in turn, and emit more neutrons. A nuclear chain reaction of rapidly increasing intensity is thus started and will continue until all the plutonium atoms are used up or until the neutrons are prevented from reacting with the plutonium atoms by being trapped in some other material or by escaping.

During the fissioning process an enormous amount of energy in the form of heat is released. This energy is equivalent to the difference between the mass of the reactants and the mass of the products of the reaction. In this instance the reactants are the plutonium atom and a neutron, and the products are the two newly produced elements of lower atomic number and the newly produced neutrons. Since the mass of the products is always smaller than the mass of the reactants in such a reaction, it would seem that some mass is lost. This difference in mass is not truly lost, however, but is changed from mass into energy according to the Einstein mass-energy equation,

$$E = mc^2$$

where E is energy, m is mass, and c is the velocity of light. (For more information about fission and nuclear structure, see *Our Atomic World,* a companion booklet in this series.)

ENERGY COMPARISONS

The nuclear energy released from one pound of plutonium is equivalent to the chemical energy that can be obtained from 3,000,000 pounds of coal (enough to fill 25 railroad cars).

We may substitute actual values for the general terms indicated by the letters in the equation to calculate the amount of energy that is equivalent to some small amount of matter. For example, if we choose 1 gram ($\frac{1}{28}$ of an ounce) for m, then for c we will use 3×10^{10} (3 followed by 10 zeros), the velocity of light in centimeters per second. In consistent units, energy will then be expressed in ergs:*

$$\text{Energy (in ergs)} = 1 \text{ gram} \times (3 \times 10^{10})^2$$
$$= 1 \times 9 \times 10^{20}$$
$$\text{Energy} = 9 \times 10^{20} \text{ ergs}$$

That amount of energy is sufficient to heat the water in a lake 1 mile long, 50 yards wide, and 13 feet deep from 70°F to the boiling point!

When nuclear energy is released quickly, an atomic explosion results. When the energy is released more slowly, as in a nuclear reactor, it can be controlled, and the resulting heat made to perform useful work. The nuclear energy released by the fissioning of 1 pound of plutonium is equivalent to the chemical energy obtained from about 20 million pounds of TNT or 3 million pounds of coal.

DISCOVERY

Pointing the Way

Between 1900 and 1934, physicists speculated a great deal about the existence of transuranium elements, and a number of scientists interpreted experimental evidence to mean that such elements could not exist. In 1934, however, Enrico Fermi, E. Amaldi, O. D'Agostino, Franco Rasetti, and Emilio Segrè, scientists at the University of Rome, discovered that neutron-irradiated uranium produced a number of radioactive substances which they suggested were "transuranium" elements. In their experiments, a glass tube filled with radon gas and beryllium powder provided

---

*An erg is a measure of the work done when a force of 1 dyne acts through a distance of 1 centimeter. A dyne is the force that will accelerate a mass of 1 gram at the rate of 1 centimeter per second per second.

neutrons, at the rate of about a million per second, to irradiate a uranium solution.

During the next 4 years, many persons studied the supposed transuranium elements. In 1938, Otto Hahn and Fritz Strassmann, Berlin scientists, succeeded in producing nuclear fission in uranium. Lise Meitner's and Otto Frisch's interpretation of that work made it apparent that most of the "new" elements really had been radioactive isotopes of elements roughly half the weight of uranium. But later, after the discovery of neptunium, it was realized that some of the products of Fermi's tests, and of the Hahn-Strassmann experiments, had been transuranium elements after all.

The Discovery

This sequel to Fermi's work stimulated much additional study of the fission process and fission products. A result was that, in 1940, Edwin M. McMillan and Philip H. Abelson of the University of California at Berkeley discovered element 93, neptunium, the first of the transuranium elements. A few months later, Arthur C. Wahl, Glenn T. Seaborg, and Joseph W. Kennedy, who had been working on the chemical properties of neptunium, also at the University of California, discovered element 94, the second transuranium element. In their experiments, uranium oxide was bombarded with neutrons in a cyclotron. (For a discussion of cyclotrons and similar equipment, see *Accelerators*, another booklet in this series.) The element 93 fraction thus formed was

Glenn T. Seaborg & Edwin M. McMillan

Joseph W. Kennedy & Arthur C. Wahl

separated chemically from the remaining oxide and examined. An alpha-emitting radioactive substance was found forming in it. Chemical examination showed that the new substance was different from element 93, and on the night of February 23, 1941, it was positively identified as an isotope (plutonium-238) of element 94.

Naming the New Element

The name plutonium was suggested by Seaborg and Wahl in 1942 in the first report ever written about the chemical properties of element 94. The two scientists thought it would be appropriate to name the second transuranium element after Pluto, the second planet in our solar system beyond Uranus, for which uranium had been named. McMillan had suggested that element 93, the first transuranium element, be named neptunium after Neptune, the first planet beyond Uranus. The suggestion was accepted; plutonium became the official name of element 94, and Pu became its chemical symbol.

EARLY RESEARCH

Because of the implication that atomic energy might be harnessed for military purposes—World War II had then started—the highly important discovery of plutonium was not announced publicly, and further work on the chemical properties of neptunium and plutonium was continued in strict secrecy. The search for plutonium-239 succeeded, confirming theoretical predictions, and on March 28, 1941, Kennedy, Segrè, Wahl, and Seaborg working at the University of California first demonstrated that plutonium-239 could undergo fission with slow (thermal) neutrons. The potential value of that isotope as a source of nuclear energy was proved. This demonstration showed that neutrons ejected in the fission process could initiate further fission and a nuclear chain reaction could be sustained. The establishment of the project for the production of plutonium metal on a large scale for possible use in a nuclear weapon quickly followed.

This highly magnified plutonium compound (2.77 micrograms of oxide) was the first to be weighed by man (September 10, 1942) and is here shown on a platinum weighing boat. The picture is magnified approximately 20 fold. The plutonium oxide appears as a crusty deposit (lower left hand part of the photograph) near the end of the platinum weighing boat, which is held with forceps that grip a small handle (upper right hand part of photograph).

Very small quantities of plutonium had been made by bombarding uranium with neutrons in a cyclotron, and it had been found possible to make microgram* amounts by stacking uranium near the cyclotron target. The neutrons captured by the uranium-238 atoms caused the formation of uranium-239, which then decayed through a two-step process to form plutonium-239. The cyclotron method, however, was incapable of producing the large quantities of plutonium that were desired.

The First Plutonium Metal

Until a solution to the production problem was found, it was necessary to use ultramicroscale methods to study the chemistry of plutonium. B. B. Cunningham and Michael Cefola, in experiments at the University of Chicago, constructed a balance sensitive to weights as small as 0.01 microgram and having a load capacity of 0.5 microgram.

*One microgram = 0.000001 gram.

By August 1942, about 50 micrograms of plutonium had been produced in compound form. In November 1943, some plutonium trifluoride (PuF_3), with some barium metal as a reductant, was heated to 1400°C in vacuum, and several plutonium metal globules, each weighing about 3 micrograms, were formed. This was the first plutonium metal.

CONTINUED STUDIES

Reactor Plutonium

The problem of producing plutonium in useful amounts was solved on December 2, 1942, when Fermi and his co-workers achieved a self-sustaining nuclear chain reaction

The front face of a plutonium producing K reactor at the Atomic Energy Commission's Hanford plant near Richland, Washington.

in a "pile" or reactor of graphite and uranium at the University of Chicago (see page 12). Within a few months, plutonium-producing reactors were built at Oak Ridge, Tennessee, and at Richland, Washington, and the secret Project Y, to build the atomic bomb, was established at Los Alamos, New Mexico. Plutonium production reactors built later at the Savannah River Plant near Aiken, South Carolina, and the reactors at Richland, Washington, have been the principal sources of plutonium in the United States. (For information about reactors and how they work, see *Nuclear Reactors*, a companion booklet in this series.)

Plutonium Laboratories

A great many university, industrial, government, and private laboratories in the United States have been studying plutonium. Laboratories from which much important information has come are operated by universities and industrial corporations for the Atomic Energy Commission.

Outside this country, information about plutonium has been produced in Canada, the United Kingdom, Russia, France, Germany, and other countries.

FORMATION

Nuclear Reaction

As mentioned earlier, neutrons captured by uranium-238 atoms cause the formation of uranium-239, which then decays radioactively to form plutonium-239. We should take a look at these processes to understand more clearly how plutonium is formed.

Fissionable Uranium-235 Normal or natural uranium consists mainly of the uranium-238 isotope with about 0.7% uranium-235 and 0.005% uranium-234. Uranium-235 is fissionable by both slow (thermal) neutrons and fast (high energy) neutrons, but uranium-238 and uranium-234 are fissionable only by fast neutrons. Since uranium-234 is present in such small amounts and is not fissionable by slow neutrons, its presence can be ignored. The uranium-235 isotope, on the other hand, is plentiful enough to be important.

The Pile Reactions If natural uranium is placed in a reactor, some of the excess neutrons produced during the fissioning of uranium-235 are captured in uranium-238 to form uranium-239. Uranium-239 has the relatively short half-life of 23.5 minutes. Each atom of it decays to form neptunium-239 by emitting a beta particle (a negatively charged electron). Neptunium-239 has a half-life of 2.35 days, and its nuclei also decay by beta emission to form

THE "PILE REACTIONS"

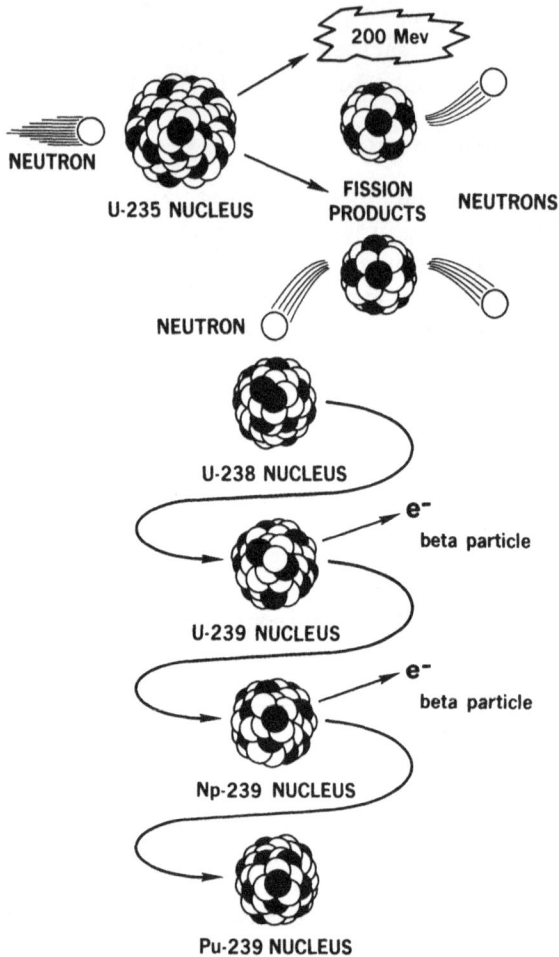

200 Mev

NEUTRON

U-235 NUCLEUS

FISSION PRODUCTS

NEUTRONS

NEUTRON

U-238 NUCLEUS

e⁻

beta particle

U-239 NUCLEUS

e⁻

beta particle

Np-239 NUCLEUS

Pu-239 NUCLEUS

When a uranium-235 atom captures a neutron in its nucleus, it fissions to produce two radioactive nuclei of fission-product elements, an average of 2.5 neutrons, and 200 Mev of energy. Although the energy produced is important, the production of more than one neutron for the one captured is even more important because that is what makes possible the rapid buildup of a chain reaction. The radioactive uranium-239 nucleus, formed when a uranium-238 nucleus captures a neutron, decays with a half-life of 23.5 minutes by emitting a beta particle (essentially an electron) to form neptunium-239 having a half-life of 2.35 days. In turn, the neptunium-239 nucleus also emits a beta particle, and is thus converted to long-lived plutonium-239.

plutonium-239. Note that the capture of a neutron by an atom increases the atomic mass of the atom by one unit but does not change its atomic number (the charge on its nucleus), and that the emission of a beta particle increases the atomic number by one unit but does not change the atomic mass. These nuclear reactions are known as the "pile reactions" and are represented by the following equations:

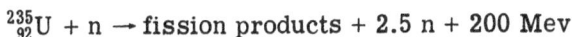

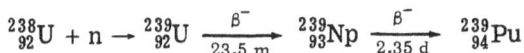

$$^{235}_{92}U + n \rightarrow \text{fission products} + 2.5\ n + 200\ \text{Mev}$$

$$^{238}_{92}U + n \rightarrow\ ^{239}_{92}U \xrightarrow[23.5\ m]{\beta^-}\ ^{239}_{93}Np \xrightarrow[2.35\ d]{\beta^-}\ ^{239}_{94}Pu$$

In these equations the supercripts are atomic masses, the subscripts are atomic numbers, n is the symbol for a neutron, m means minutes, d means days, and Mev means million electron volts (a convenient unit of energy used in physics).

The equations show that about one atom of plutonium-239 can be formed per atom of uranium-235 undergoing fission and that fission products are formed along with the plutonium. The fission products include highly radioactive isotopes of several elements. Their presence complicates the separation and recovery of plutonium from uranium since the operation must be performed remotely from behind thick shielding to protect the technicians from the radiation.

Natural Occurrence

Although for all practical purposes plutonium must be made synthetically, careful analysis has shown that plutonium does occur naturally in some uranium ores. C. A. Levine and G. T. Seaborg, after examining Belgian Congo pitchblende concentrate (uranium content of 45.3%), reported in 1951 that it contained about 7 parts of plutonium-239 in a trillion (1×10^{12}) parts of the concentrate. Fergusonite ore (uranium content of 0.25%) contained about 1 part of plutonium-239 in 1×10^{14} parts. No isotopes of plutonium other than plutonium-239 were found. It is therefore unlikely that significant amounts of plutonium ever will be produced economically from even the richest uranium deposits.

The half-lives of the plutonium isotopes, with the possible exception of plutonium-244, which has a half-life of 76 million years, are so short compared with the age of the earth that any plutonium which existed when the earth was young could hardly still exist today. The plutonium now found in nature probably was formed by natural nuclear reactions that go on continuously. Neutrons necessary for the formation of plutonium-239 from uranium-238 may come from several sources, including (1) nuclear reactions resulting from the interaction of light elements with alpha particles emitted by uranium and its daughters, (2) spontaneous fission of uranium-235, and (3) cosmic rays.

PRODUCTION

After plutonium-239 has been formed in a reactor, it continues to be bombarded by neutrons, which convert some of the 239 isotope to plutonium-240 and some of the 240 to 241 or even higher isotopes. The amounts of higher isotopes

Plutonium Production Flow Diagram

Some of the uranium-238 in reactor fuel elements is converted to plutonium-239 as the reactor is in operation. The used fuel elements, containing plutonium, uranium, and fission products, are removed from the reactor and allowed to "cool" under water for from 2 to 4 months before being sent to the dissolver.

In the dissolver, the aluminum jackets are removed, and the plutonium, uranium, and fission products are put into solution. Chemical processing separates the plutonium and uranium for further steps and the fission products for storage and disposal.

The plutonium, usually in the form of a nitrate solution, goes through a precipitation step where it is converted to a peroxide. This is treated with hydrogen fluoride (HF) to produce the compound, plutonium tetrafluoride (PuF₄).

The PuF₄ is then reduced to plutonium metal in a thermal process that uses calcium and iodine. The metal "buttons" resulting from this process are cleaned by "pickling" before they are melted, cast, and machined or otherwise formed into a finished item.

All residues and scrap from the operations that follow chemical processing are sent through a recovery operation to retain all significant amounts of plutonium in the system.

produced with the plutonium-239 depend on the type of reactor and how it is operated. They rarely make up more than a few percent of the total plutonium.

Formation of the element, however, is only the first step toward producing the metal. Next, the plutonium must be separated from the uranium and the fission products. It is fortunate that the chemistry involved in this separation is straightforward and quite simple because the process must be performed under two very difficult conditions: (1) very small amounts (of the order of grams) of plutonium must

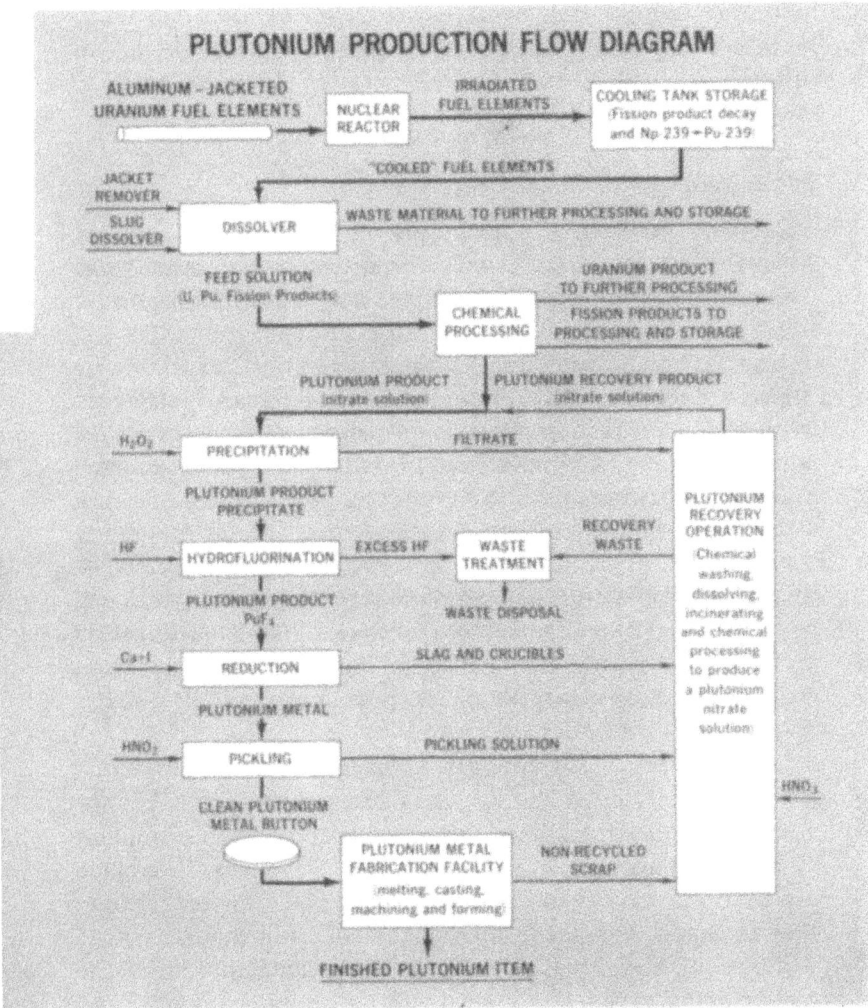

PLUTONIUM PRODUCTION FLOW DIAGRAM

be separated from tons of uranium, and (2) the people doing the work must be protected against the intensely radioactive fission products.

Cooling Period

When the uranium-238 in the reactor fuel has been irradiated for a sufficient time to form an appreciable amount of plutonium-239, the irradiated fuel elements are removed from the reactor and stored under water for 2 to 4 months. During this storage, called a cooling period, the highly radioactive fission products decay to lower levels of activity. In this time, also, most of the neptunium-239, which was initially formed from the uranium-238, is transformed to plutonium-239.

Separation

Chemical Methods   After the cooling period, any of several methods can be used to separate the plutonium from the uranium and fission products. The chemical methods, which are generally used in large scale operations, depend on the fact that plutonium can exist in several oxidation or valence states and that its chemical properties are different from those of uranium in one or more of the states. It is possible to prepare solutions in which the plutonium and uranium are present in different oxidation states and have different chemical properties. Separation of the plutonium by selective precipitation or solvent extraction methods is then relatively easy. In a similar manner the plutonium can be separated from the fission products. Other separation methods, including ion exchange and pyrometallurgical (high-temperature metallurgical) processes, can be used in special circumstances.

Ion Exchange   In the ion exchange method, plutonium is adsorbed from dilute aqueous solutions by ion exchange resins to separate it from the uranium and fission products.

Pyrometallurgy   Pyrometallurgical processes include distillation, molten-metal extraction, and salt extraction. Use of these processes avoids the need for handling large volumes of solutions, but has the disadvantage of having to be performed at high temperatures.

16

The distillation method is useful because of the large difference between the vapor pressures of plutonium and uranium at high temperatures. At 1540°C the vapor pressure of plutonium is 300 times that of uranium, and it therefore can be boiled off or distilled from molten uranium.

Some metals, such as silver or magnesium, when molten have the property of dissolving considerable amounts of plutonium but very little uranium. The dissolving metals then can be removed from the plutonium by distillation.

In the salt extraction method, the plutonium may be absorbed in a salt to separate it from the uranium. In a molten mixture of uranium tetrafluoride (UF_4) and irradiated uranium metal, 90% of the plutonium in the metal goes into the salt. If the irradiated metal is repeatedly exposed to fresh salt, eventually almost all the plutonium can be recovered.

Reduction to Metal

The product of most of these separation methods is a plutonium salt which must be processed to obtain the element in metallic form. An example of such a process is the reduction of plutonium tetrafluoride (PuF_4) by heating the PuF_4 with a reducing agent, such as calcium, in a pressure chamber (bomb reduction process). The purity of the metallic plutonium produced by this method is normally about

The product from the thermal reduction process is a solid chunk or "button" of plutonium metal. The button shown in the photograph weighed about $3/_4$ pound and was roughly 3 inches in diameter.

99.87%. If plutonium of exceptional purity is wanted, it can be produced by electrorefining the metal obtained from the bomb reduction method. An impure plutonium metal anode is used in an electrolyte solution containing plutonium chloride, potassium chloride, and sodium chloride. A metal tube of either tantalum or tungsten is the cathode completing the electrolytic cell. This cell is operated at 700°C in a helium atmosphere. At this temperature the impure plutonium is oxidized to form trivalent ions which enter the molten salt electrolyte and are transported through it to the cathode, where they are reduced back to metal at the cathode surface. Since the operation is performed at a temperature above the melting point of plutonium, the purified molten plutonium can be continuously drawn off as a liquid from the cathode. Plutonium made by this method may be 99.98 or 99.99% pure.

Fabrication

Melting Plutonium metal must be melted in vacuum or in an inert atmosphere (helium or argon) because the metal is very reactive in air at high temperatures. Also, plutonium has strong reducing properties that allow only the most stable oxide, carbide, nitride, boride, and silicide compounds to be used as crucible materials. Crucibles made of oxides of magnesium (magnesia) and calcium (calcia), which have been high-fired (heated to high temperature before use), can be used to contain molten plutonium at temperatures below about 1200°C. Oxides of yttrium (yttria) and thorium (thoria) have been used as crucible materials at high temperatures (to about 1500°C) with some success, and crucibles made of very high melting-point metals, such as tantalum or tungsten, can be used if the temperatures do not exceed about 1000°C. Steel crucibles that have been chromium plated and then coated with a layer of calcium fluoride are also used.

Casting Molds of graphite or metal (cast iron or mild steel) are often used in casting plutonium. The inside surfaces of the molds are coated with calcium fluoride to prevent the molten plutonium from sticking to the mold or reacting with it.

Forming   Plutonium, like most solids, is made up of many small, cohering crystals or grains, and the pure metal is normally fine-grained, hard, and brittle. It is about as hard and brittle as gray cast iron, and it can be machined like cast iron.

A common plutonium alloy containing 1% aluminum is about as soft as annealed copper. This alloy is so ductile that it can be formed into rods, wires, sheets, or "dished" shapes by such fabrication methods as extruding, drawing, rolling, and spinning.

In all these operations conventional tools and equipment are used, but precautions must be taken to protect the operators from exposure to plutonium contamination. These precautions are a highly important aspect of handling plutonium and will be discussed later.

PROPERTIES

Plutonium has been known to be in existence for only two decades. It is produced and handled only with considerable difficulty, and much of the research on the metal during the past 20 years has necessarily been concerned with developing production methods rather than the metal's properties. Nevertheless, an impressive amount of information about its properties has been collected. Many details remain to be explained before we have a good understanding of plutonium, and much of our present information is not complete. Very small amounts of impurities may have a surprisingly large effect on the properties of metals. For example, the presence of as little as 0.0005% bismuth in copper causes the copper to become slightly embrittled at temperatures near 600°C; copper containing between 0.005 and 0.01% bismuth is entirely unsuitable for hot-rolling operations. Thus it is conceivable that the 0.01-0.02% impurities in the purest plutonium yet available may have a considerable effect on the properties.

Some of the accepted information about the nuclear, chemical, and physical properties of plutonium is given in this section. Additional information and details may be found in the references on page 51.

Nuclear

The nuclear reactions that are responsible for the formation of plutonium were mentioned previously, but its radioactive decay properties were ignored. Now we will learn what happens to plutonium after it has been formed.

Plutonium-239 is not stable. As soon as it is formed, it begins to change into uranium-235, the first step in its decay process, by ejecting alpha particles. (An alpha particle is the same as a helium nucleus, with an atomic mass of 4 units and an atomic number of 2. Therefore, when an alpha particle is ejected, the daughter element formed must be 4 atomic units and 2 atomic number units less than the parent. That is, $^{239}_{94}Pu \xrightarrow{He^{++}} {}^{235}_{92}U$). Plutonium-239 has a half-life of 24,360 years, and it takes many years for a significant amount of the plutonium to be transformed to uranium-235.

THE FIRST STEP IN THE RADIOACTIVE DECAY OF PLUTONIUM

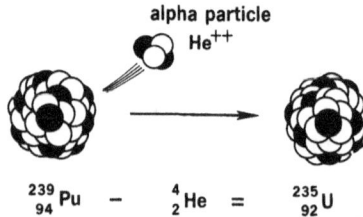

$$^{239}_{94}Pu - {}^{4}_{2}He = {}^{235}_{92}U$$

As the first step in its radioactive decay chain, the plutonium-239 nucleus emits an alpha particle (essentially a helium nucleus of atomic weight 4 and atomic number 2) and is thus transmuted into a uranium-235 nucleus. Plus signs, $^{++}$, indicate positive charges on the nucleus.

Uranium-235 also decays radioactively by alpha emission but with the much longer half-life of more than 700 million years. It forms thorium-231. Eventually, after passing through a total of 15 radioactive decay steps, some involving elements having half-lives as short as a few thousandths of a second, plutonium-239 becomes the stable (nonradioactive) element lead.

Other plutonium isotopes also decay, some by different sequences. For example, plutonium-241 changes first to americium-241 by emitting a beta particle and then continues to change through 13 more steps to become stable bismuth.

The alpha particles given off by plutonium do not travel very far. Their range in air is less than 1.5 inches, and they will not penetrate a sheet of paper. However, when an alpha particle is ejected from the plutonium-239 atom, the resulting uranium-235 nucleus recoils with considerable energy. Both the recoiling uranium-235 nucleus and the alpha particle are responsible for displacing atoms and damaging the crystal structure of the metal.

The alpha activity also results in self-heating in the metal. The self-heating coefficient (the rate at which the heat is generated) is about 0.0005 calorie per gram per second. Metal containers holding large specimens of plutonium (1 or 2 pounds) feel warm when touched with the hand.

Chemical

Reaction in Gases   Plutonium is a highly reactive metal. It readily combines with oxygen to form plutonium dioxide (PuO_2) or, under special conditions, the monoxide (PuO) or sesquioxide (Pu_2O_3). In dry air the oxidation reaction proceeds rather slowly because the oxide coating formed on the metal surface protects the underlying metal. The surface oxide formed in ordinary humid air, however, is too powdery to protect the metal.

Effect of Temperature   Increasing the temperature of plutonium exposed to ordinary air rapidly increases the oxidation rate. Therefore the metal must be protected in some manner when it is heated. Usually a good vacuum or a dry inert-gas atmosphere (helium or argon) provides sufficient protection, but even under the best conditions some oxide will form.

At high temperatures all the common gases (carbon monoxide, carbon dioxide, ammonia, nitrogen, hydrogen, fluorine, and chlorine) react with plutonium.

Reactions with Acids and Bases   Plutonium is readily dissolved by concentrated hydrochloric, hydroiodic, and perchloric acids. Most dilute acids will attack it. Concentrated sulfuric and nitric acids and aqueous bases such as sodium hydroxide solutions do not attack the metal readily, however.

Compound Formation   Plutonium has four oxidation or valence states, usually noted as (III), (IV), (V), and (VI), and forms quite stable compounds with all the nonmetallic elements except the rare gases. Because of their stability, such compounds as plutonium dioxide, plutonium carbide, and plutonium nitride are being studied to determine their potential value as nuclear fuels.

Physical

Allotropes   Perhaps the most unusual physical property of plutonium is its occurrence in six different crystal structure forms or allotropes, each in a certain well-defined temperature range. It is not uncommon for elements to have more than one allotrope, but plutonium is the only one with as many as six. At room temperature, plutonium normally exists in what is called the alpha (α) phase. When heated to about 115°C, its crystal structure undergoes a rearrangement of atoms to form a different crystal structure, the beta (β) phase. Further heating to 185, 310, 452, and 480°C results in gamma (γ), delta (δ), delta-prime (δ'), and epsilon (ϵ) phases, respectively.

Table 1
Some Properties of the Plutonium Allotropes

Allo-trope	Crystal Structure	Density	Linear Thermal Expansion Coefficient for Temperature Range Given	
α	Simple mono-clinic	19.86 g/cm³	$+54 \times 10^{-6}$/°C	(21–104°C)
β	Body-centered monoclinic	17.70	+42	(93–190°C)
γ	Face-centered orthorhombic	17.14	+34.6	(210–310°C)
δ	Face-centered cubic	15.92	−8.6	(320–440°C)
δ'	Body-centered tetragonal	16.00	−65.6	(452–480°C)
ϵ	Body-centered cubic	16.51	+36.5	(490–550°C)

Thermal Expansion   In the alpha, beta, gamma, and epsilon phases, plutonium expands while being heated—as do most metals—but in the delta and delta-prime phases it contracts. This unusual contraction has been the subject of much study.

The rates at which plutonium expands while being heated (except in the delta and delta-prime phases) are unusually high. At about 20°C only lithium, sodium, potassium, rubidium, and cesium have higher expansion rates. Some properties of the plutonium allotropes are listed in Table 1, and some physical constants of the metal are given in Table 2.

Density   The densities of the plutonium allotropes are high (see Table 1), ranging from 15.92 g/cm^3 (grams per cubic centimeter) for the lowest density phase (delta) to 19.86 g/cm^3 for the highest density phase (alpha). Only iridium, neptunium, osmium, and platinum have higher densities than alpha plutonium. Lead, often thought of as a very dense metal, has a density of only 11.34 g/cm^3. Gold at 19.32 g/cm^3 is slightly less dense than alpha plutonium.

Table 2*
Some Physical Constants of Plutonium

Atomic number	94
Isotopic mass, plutonium-239 (chemical scale)	239.06
Atomic weight	239.11†
Melting point	640°C
Boiling point	3327°C
Vapor pressure (1120–1520°C)	$Log_{10} P_{mm} = -\dfrac{17,420}{t + 273.18} + 7.794$
Average heat of vaporization (1120–1520°C)	79.7 kcal/g-atom

*Courtesy Reinhold Publishing Corporation, *Rare Metals Handbook*, 1961.
†Computed atomic weight (chemical scale) of plutonium containing 95.37 atomic percent plutonium-239, 4.43 percent plutonium-240, and 0.20 percent plutonium-241.

EXPANSION BEHAVIOR OF PLUTONIUM

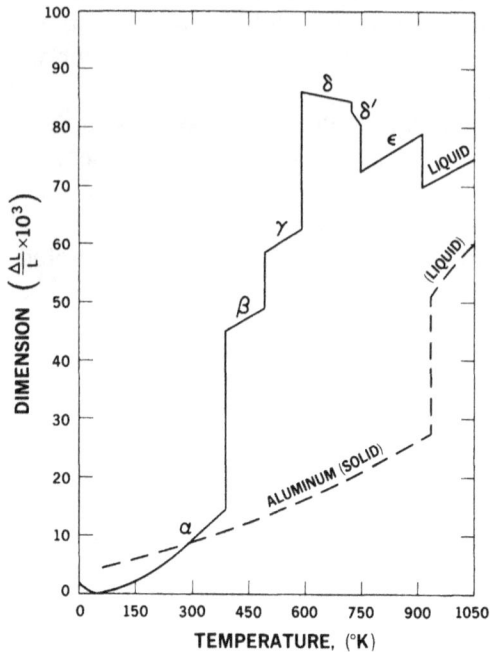

Idealized version of the expansion behavior of plutonium, showing approximate temperature ranges in which the allotropes or phases of plutonium normally exist. The effect of temperature on the dimensions of a plutonium specimen is also shown.

When plutonium is heated from a temperature near absolute zero ($0°K$ or $-273°C$), it is in the alpha phase and shrinks slightly until $50°K$ ($-223°C$) is reached, then expands with further heating to $388°K$ ($115°C$). At this temperature the specimen transforms to the beta phase undergoing marked expansion at constant temperature.

With heating to $583°K$ ($310°C$), the specimen expands in the beta phase, during beta-to-gamma transition, in the gamma phase, and during gamma-to-delta transition. In the delta phase, however, the specimen shrinks while being heated. It also shrinks as it transforms to the delta-prime phase at $725°K$ ($425°C$), as it is heated in the delta-prime phase, and as it transforms to the epsilon phase at $753°K$ ($480°C$).

Above that temperature the specimen is in the epsilon phase, and expands during heating until the melting point is reached at $913°K$ ($640°C$). The specimen shrinks as it changes from solid to liquid, then expands as it is heated in the liquid state.

The dashed line represents the simpler, more normal expansion of aluminum.

Melting Point   Plutonium's melting point, 640°C, is quite low for a metal. For comparison, iron melts at 1535°C. Tungsten has the highest melting point of all metals, 3410°C.

Boiling Point   The boiling point of plutonium is exceptionally high, 3327°C, considering its low melting point. For comparison, iron boils at 3000°C, tin at 2270°C, and tantalum, with the highest boiling point of all the elements, at 6100°C.

Thermal Conductivity   The thermal conductivity (ability to conduct heat) of plutonium is low. Values of 0.02 and 0.008 cal/cm^2/cm/sec/°C have been reported by different scientists. The reason for this discrepancy is not known, but it is certain that the true value is low for a metal. Of

Table 3*
The Specific Heat of Plutonium

Phase	Temperature °C	Specific Heat Cal/g-atom
α	30	8.6
	60	8.6
	90	9.0
β	140	8.3
	190	8.4
γ	220	8.6
	250	8.8
	300	9.6
δ	330	9.0
	420	9.0
δ'	No true values	
ε	490	8.4
	550	8.4
	600	8.4
	650	9.9
Liquid	660	10.0
	675	10.0

*Courtesy Cleaver-Hume Press, Ltd., *Plutonium* 1960, 1961.

all metals, silver has the highest thermal conductivity, 1 cal/cm^2/cm/sec/°C. Fireclay, a thermal insulating material, has a thermal conductivity of about 0.001 cal/cm^2/cm/sec/°C.

Electrical Resistivity   Most metals have low electrical resistivities. The resistance of copper, for example, is only about 10 microhm-cm at 1000°C and decreases quite uniformly to zero at absolute zero (−273°C). That of plutonium, however, is unusually high. Among the metals only manganese has greater electrical resistivity. At room temperature the electrical resistivity of plutonium is about 145 microhm-cm, higher than that of bismuth (about 120 microhm-cm) but lower than that of manganese (about 188 microhm-cm). The resistivity of plutonium is at its maximum at about 157 microhm-cm near −173°C.

Specific Heat   The specific heats of the plutonium allotropes are given in Table 3, and the heats of transformation are listed in Table 4.

Heat of Fusion   The heat absorbed by plutonium during melting is reported to be 676 ± 10 cal/g-atom, which is lower than most of the known values of this property for other metals.

Alloy Formation   Plutonium forms alloys and intermetallic compounds with most of the metals. Exceptions are lithium, sodium, and potassium (and perhaps rubidium

Table 4
Transformation Temperatures and Heats of Transformation of the Plutonium Allotropes

Transformation	Transformation Temperature (°C)	ΔH* (Cal/g-atom)	ΔS† (Cal/deg/g-atom)
α → β	115	900 ± 20	2.32
β → γ	≈185	160 ± 10	0.35
γ → δ	310	148 ± 15	0.25
δ → δ′	452	10 ± 10	0.01
δ′ → ε	480	444 ± 10	0.59
ε → liquid	640	676 ± 10	0.74

*Amount of heat absorbed by changing from one allotrope to another.
†Increase in entropy, the energy unavailable for work.

and cesium) of the alkali metals; calcium, strontium, and barium of the alkaline-earth metals; and europium and ytterbium of the rare-earth metals, all of which are immiscible with both solid and liquid plutonium. Tungsten, tantalum, molybdenum, niobium, chromium, and vanadium are soluble in liquid plutonium but are insoluble or only slightly soluble in solid plutonium. Iron, cobalt, and nickel form low-melting alloys with plutonium. The plutonium alloy containing about 10 atomic percent* iron melts at 410°C, and plutonium alloys containing about 12 atomic percent cobalt or about 12 atomic percent nickel melt at about 408 and 465°C, respectively. (See cover.)

In summary, although plutonium must be termed a metal, it does not have such usual metallic properties as high electrical and thermal conductivities and reasonably good ductility. To a metallurgist working with it or a scientist studying its properties, plutonium is a complicated material because of the many different crystal structures it passes through as it is heated and because many of its properties are so different from those of more common metals. These properties of plutonium are complicated further by its radioactivity, which causes it to self-heat, to damage itself, and to change in chemical composition as the radioactive decay products accumulate.

USES

The first atomic bomb, exploded in the desert 60 miles northwest of Alamogordo, New Mexico, on July 16, 1945, contained plutonium. It was with this explosion that the reality of the atomic age became clear.

Nuclear Reactors

Plutonium-fueled Reactors   In November 1946, 16 months after the horrible effectiveness of plutonium as an instrument of destruction had been demonstrated, Clementine, the first nuclear reactor to use plutonium as fuel, was started up at the Los Alamos Scientific Laboratory in New

---

*Atomic percent gives the number of atoms, in proportion to one hundred, rather than units of mass or volume.

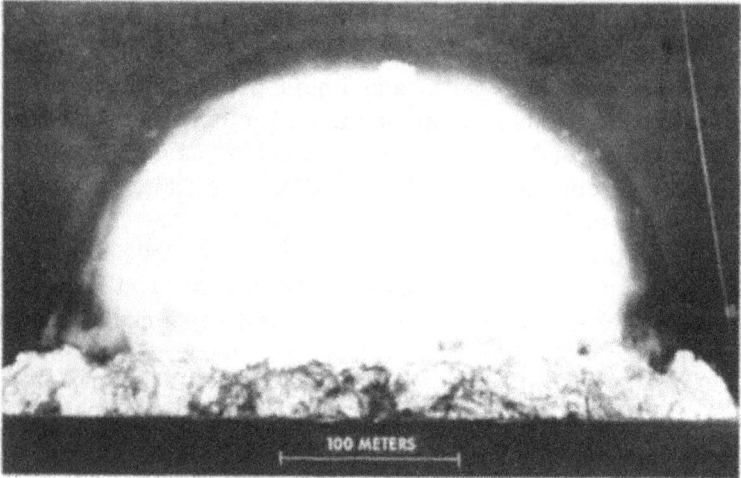

The fireball of the Trinity explosion 0.053 sec after detonation, as it shook the desert near the town of Alamogordo, New Mexico, on July 16, 1945. This atomic device, which marked the first use of plutonium, was developed and built at the University of California's Los Alamos Scientific Laboratory in New Mexico.

Mexico. Clementine was a mercury-cooled research reactor and was operated until 1953.

Since 1953 two research reactors in Russia, BR-2 and BR-5, and three in the United States, the Materials Testing Reactor (MTR), the Los Alamos Molten Plutonium Reactor Experiment No. 1 (LAMPRE-1), and the Experimental Breeder Reactor No. 1 (EBR-1), have operated with plutonium cores. Canada's NRX reactor has used some plutonium-bearing fuel elements in conjunction with uranium fuel elements. The Experimental Breeder Reactor 2 (EBR-2) and the Plutonium Recycle Test Reactor (PRTR) have been constructed to establish the feasibility of using plutonium in power reactors. (See photos, pages 30, 31, 32.)

Importance of Plutonium as a Reactor Fuel It has been apparent for a number of years that plutonium must be further developed as a reactor fuel if our nuclear energy resources are to be used economically and efficiently. Normal uranium contains about 140 times as much uranium-238 as uranium-235, but only the 235 is readily fissionable. The uranium-238, however, is fertile. It can be converted into plutonium, an easily fissionable material. Therefore, if the plutonium formed in reactors by neutron irradiation of uranium-238 were also used as a reactor fuel, our nuclear fuel reserves could be increased by several orders of magnitude.

Unfortunately, however, unalloyed plutonium has several properties that are undesirable in a nuclear fuel: (1) Its complicated expansion behavior probably would change the shape and dimensions of solid fuel elements so drastically that they would be useless. (2) The thermal conductivity and melting point of plutonium are so low that the temperature at the surface of a fuel element would have to be kept low (a disadvantage in a reactor) if melting were to be avoided at the center of the element. (3) The high rate of heat generation combined with the low thermal conductivity of both solid and liquid plutonium severely limits the size of a reactor core from which the heat can be extracted efficiently. In view of these disadvantages, it appears that plutonium must be alloyed or mixed with other metals or elements if it is to be used efficiently in large power reactor systems. Indeed, considerable attention is being paid to

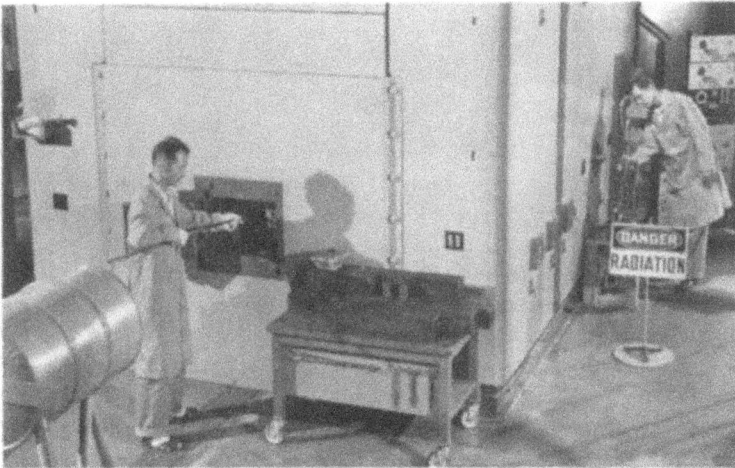

Los Alamos Fast Reactor, Clementine, was the world's first reactor to use fast neutrons and plutonium. The scientist in the foreground of the photograph is inserting some material into the reactor to be irradiated by neutrons.

plutonium alloys and plutonium oxides, carbides, and nitrides (ceramic fuels) in an effort to develop satisfactory plutonium-containing fuels.

Phoenix Fuels　　An interesting material containing certain combinations of the plutonium isotopes 239, 240, and 241 is also being studied. These combinations, which are expected to make possible extraordinarily long-lived reactor cores, are called Phoenix Fuels.* As the plutonium-239 is burned up in the reactor, the fertile plutonium-240 isotope yields enough fissile plutonium-241 to maintain the reactivity of the fuel.

Aerial view of the Experimental Breeder Reactor (EBR-2 complex at the National Reactor Testing Station, Idaho. The laboratory and service building (1) is located in the foreground. The reactor and the primary sodium coolant system are in the reactor plant (2), within a steel containment vessel 80 feet in diameter by 140 feet high. The power plant (3) contains the control room for the reactor and the steam and electrical equipment. Facilities for processing and fabricating the fuel elements are in the fuel cycle facility (4), and the sodium-to-water heat-transfer equipment to generate the steam is in the sodium boiler plant (5).

---

*The phoenix was a mythical bird in Egyptian religion. When consumed by fire it would arise again from its own ashes.

PLUTONIUM FUEL TECHNOLOGY

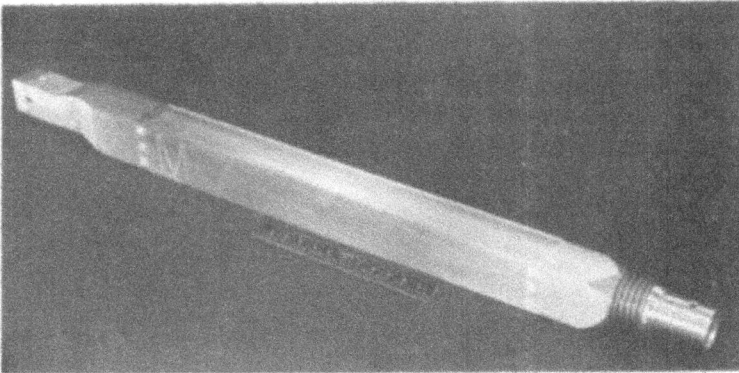

A fuel element of the type used in the Materials Testing Reactor (MTR). The assembly consists of a number of aluminum-clad plutonium—aluminum alloy plates that have been brazed together to form the complete element shown in the photograph.

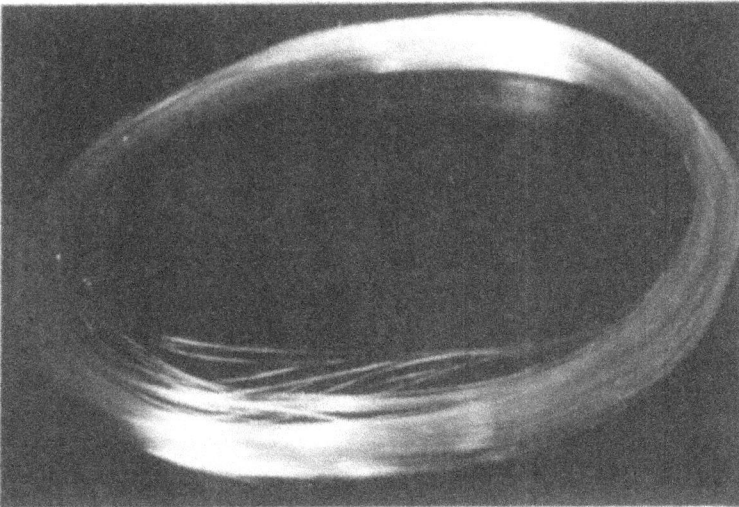

Plutonium oxide has been successfully incorporated into silicate glass formulations. The plutonium oxide is so tightly bound in the glass matrix that danger from handling these glasses has been eliminated. A glass may be fabricated into various shapes and forms. The coil of glass fibers in the picture was drawn by the monofilament technique from a glass billet containing ten weight percent plutonium oxide. These plutonium-bearing glasses developed at the AEC's Mound Laboratory are presently being evaluated as a potential reactor fuel at Argonne National Laboratory.

LAMPRE I

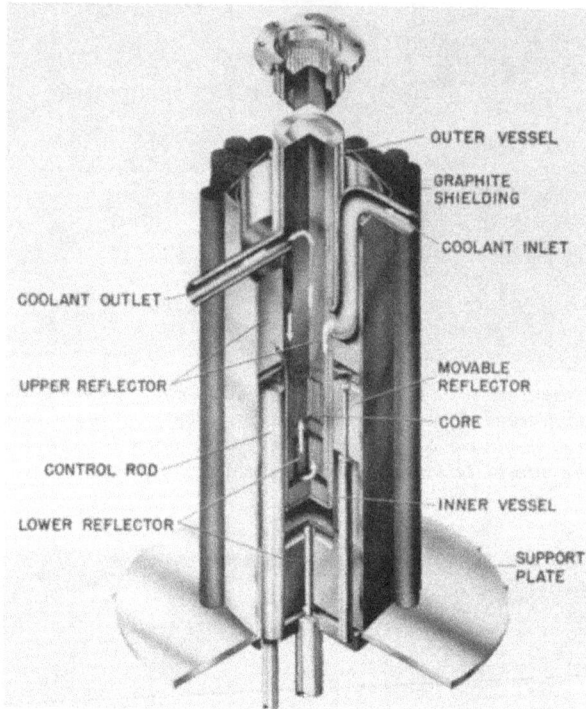

Labels on figure:
- OUTER VESSEL
- GRAPHITE SHIELDING
- COOLANT INLET
- COOLANT OUTLET
- MOVABLE REFLECTOR
- UPPER REFLECTOR
- CORE
- CONTROL ROD
- INNER VESSEL
- LOWER REFLECTOR
- SUPPORT PLATE

The core of this experimental reactor consisted of 199 tantalum capsules, each about $\frac{3}{8}$ inch diameter by 8 inches long. The capsules contained a plutonium alloy fuel that was liquid at the operating temperature of the reactor (in the vicinity of 500°C).

Isotope Power Generators

In addition to being used in nuclear weapons and reactors, plutonium has been used as the source of energy in some isotope power generators. Another isotope, plutonium-238, is used in some of these generators, which are being developed under the Systems for Nuclear Auxiliary Power (SNAP) program. (The SNAP program is described in *Power from Radioisotopes,* a companion booklet in this series.) The generators convert the heat energy from the spontaneous disintegration of a radioactive isotope directly into electrical energy. Ordinarily, the conversion of heat to electrical power requires an intermediate step: heat from burning coal produces steam, which is a source of mechanical power, to drive an electrical generator.

Thermoelectric Conversion   One direct method of converting heat to electrical power is thermoelectric conversion. If two different metals are joined in a closed circuit and the two connections where one metal joins the other are held at different temperatures, an electrical current will be produced in the circuit.

Thermionic Conversion   Thermionic emission or conversion is another type of direct energy conversion. In this method an electric current is produced in a diode vacuum (or gas-filled) tube by heating the cathode and collecting the emitted electrons on the anode. For this process, plutonium is used less than other radioisotopes, however. (For additional information on these processes see *Direct Conversion of Energy*, a companion booklet in this series.)

Isotope generators thereby provide a source of electrical energy that has no moving parts to wear out and do not require the frequent attention and maintenance necessary with battery-powered equipment. Such generators are especially useful in providing power for satellites, unattended automatic weather stations in remote locations, navigational buoys, and other equipment which cannot be visited frequently and on schedule.

Neutron Sources

Neutron sources that are useful for calibrating neutron detection equipment are made by alloying plutonium and beryllium to form the compound $PuBe_{13}$. In this compound the beryllium atoms and the alpha particles emitted by the plutonium atoms react to produce neutrons and carbon atoms; that is

$$^{9}_{4}Be + ^{4}_{2}He \rightarrow ^{12}_{6}C + ^{1}_{0}n$$

Such compounds emit a constant number of neutrons for a given weight and, if plutonium-239 is used, are long-lived and stable. A different type of source, in which neutrons are emitted during spontaneous fissioning of plutonium-240, is made from highly irradiated plutonium.

Threshold Detectors

Plutonium has been used as a threshold detector for determining neutron energies. For example, if a plutonium foil surrounded by boron-10 (an isotope of boron) is exposed to a flux of neutrons, the activity of the resulting fission fragments provides information about the intensities of neutrons having energies higher than the threshold value of 4 kilovolts.

Transplutonium Elements

Transplutonium elements and higher isotopes of plutonium for research use can be produced by irradiating plutonium in the intense neutron flux of a reactor.

Thermonuclear Explosives

Energy from the fission of plutonium may some day initiate thermonuclear (fusion) explosives for producing useful peacetime work. Large-scale canal construction and harbor excavation, for example, might be done economically and efficiently by thermonuclear explosives, in which, however, the bulk of the energy would come from the fusion reaction.

HANDLING

There are two special types of precautions that must be taken by those who work with plutonium. One is a consequence of plutonium's fissionable nature; the other, of its alpha-emitting property.

Criticality

Solid If a sufficient amount of solid metallic plutonium, of the order of a few pounds, is brought together in one place, a critical mass is formed. That is, for each neutron causing a plutonium atom to fission, another neutron is produced which causes another plutonium atom to fission. This chain reaction, accompanied by the emission of lethal amounts of neutrons and gamma rays, will continue as long as the critical mass exists. Moreover, since a large amount of heat will be generated in a short time, a reaction approaching an explosion may occur. The violence of the reaction will depend on how rapidly the critical mass is formed and how closely it is confined.

Liquid   If the plutonium is in an aqueous solution rather than a solid metal, it takes a much smaller quantity to become critical. This can be as little as about one-tenth of a solid critical mass, depending on such conditions as the volume of the solution and the shape of the container holding it. The explanation is that the hydrogen atoms in the water are highly effective in reducing the speed of neutrons and thus increase the neutrons' ability to cause fission.

Precautions

In a plutonium production plant, where large quantities of plutonium are handled, elaborate precautions are taken to prevent the unintentional formation of a critical mass. For example, in some of the processes involved in producing plutonium, the tanks, vats, reaction vessels, and other containers are of such dimensions and capacities that a critical mass cannot be formed in them under any normal condition. Batch processing, in which a small amount of plutonium (considerably less than a critical mass) is carried completely through the operation before another batch is admitted to the same operation, can also be used. Careful attention is paid continuously to the sizes and shapes of the pieces of plutonium being worked with or assembled in storage areas. Gamma radiation detection equipment is used to provide an automatic warning of the presence of a high level of gamma-ray activity, and a strict accounting system that shows the amount of plutonium on hand in the various areas of the plant at any time is in effect.

Toxicity

Radioactivity   In addition to the criticality hazard, which exists only in relatively massive amounts of plutonium, its highly toxic nature must also be considered. This is a consequence of the element's alpha-emitting property and must be reckoned with even in particles of plutonium so small as to be invisible.

The alpha radiation given off by plutonium will not penetrate the surface layer of unbroken skin, but if it gets into human tissues in other ways it can cause severe damage. This radiation is especially damaging to the blood-forming organs in the bones and can produce bone diseases even many years later.

A Plutonium Laboratory Building

The Chemistry and Metallurgy Research Building at Los Alamos Scientific Laboratory, Los Alamos, New Mexico. The right-hand wing is occupied by the Plutonium Physical Metallurgy Group. It is one of seven wings connected by a "spinal" corridor 650 feet long.

The building, which was designed specifically for work with plutonium, incorporates several features to deal with the health hazards involved in handling it. Among these is an elaborate and extensive ventilation system. The entire building is air-conditioned, windows are of transparent glass block that cannot be opened, and each wing in which plutonium is used has its own ventilating system.

Supply air enters at the attic level, is filtered and washed before being directed downward to laboratories, offices, and storage areas. Because plutonium and its compounds are very dense, plutonium-contaminated air must be exhausted downward, so all air from laboratory wings is carried out of the building through large stainless-steel ducts below the floor.

Before the exhaust air is released to the outside atmosphere, it passes through an elaborate filtering system to remove all contamination, and is continuously sampled for assays of radioactivity. The exhaust filter room is at the end of the wings; metal panels cover openings through which filtering equipment can be replaced or repaired.

It may enter the body through cuts or abrasions of the skin, by being swallowed or, most importantly, by inhalation. Once in the body, plutonium is eliminated so slowly that as much as 80% of any amount taken in will still be there 50 years later. One to ten percent is likely to be deposited in the bones.

Allowable Levels   Because of the indiscernible nature of the health hazard, a limit has been placed on the amount of plutonium which a person should accumulate in his body. This maximum permissible body burden, or the total amount of plutonium that can be accumulated in an adult without eventually producing undue risk to health, has been set at 0.6 microgram. A particle of plutonium the size of a single, ordinary dust particle weighs about that much. Correspondingly, a limit has also been placed on the amount of plutonium contamination that can be safely allowed in the air people breathe. This limit is called the "maximum allowable concentration" and is only 0.00003 microgram of plutonium per cubic meter of air.

Contamination Control   Strict control of radioactive contamination is imperative in a plutonium operation. In some laboratories all forms of plutonium are kept in hermetically sealed enclosures, the atmospheres within these enclosures never being allowed to come in contact with the atmosphere breathed by workers in the open laboratory. This type of protection is termed "total containment" and is especially suitable where operations can be performed by automated equipment.

In other laboratories the importance of convenience may outweigh the reliability of total containment, and, although most of the work is performed within glove boxes or other enclosures, a few suitably clean and protected samples of plutonium may be taken into an open laboratory. In this situation, adequate ventilation is necessary to keep plutonium contamination within the allowable level. Of course, where people have easy access to plutonium and to the equipment used in studying it, scrupulous attention must be paid to cleanliness and to maintaining all surfaces in the laboratory as free from contamination as possible. Various types of radiation-detection equipment — air samplers, proportional alpha counters, beta—gamma survey meters, neu-

A glove-box train used in preparing metallographic specimens containing plutonium. An analytical balance for weighing specimens is visible in the glove box at the left. Specimens are brought into the train through the open front hood (second enclosure from the left), and a variety of operations (cutting, filing, grinding, polishing) are performed in the boxes at the right.

Handling Plutonium

Handling hot plutonium-alloy ingots in a glove box. Three operators are required. One man (foreground) removes the hot ingots from a pot furnace in the glove box and hands them to the rolling-mill operator (right), who runs them through the rolls of the mill to reduce their thickness. The third man (left, behind box) receives the ingots as they emerge from the rolls and returns them to the operator for another pass through the mill or to the furnace for reheating. The men communicate with each other by means of throat microphones and earphones, and wear asbestos gloves to protect the glove box from damage.

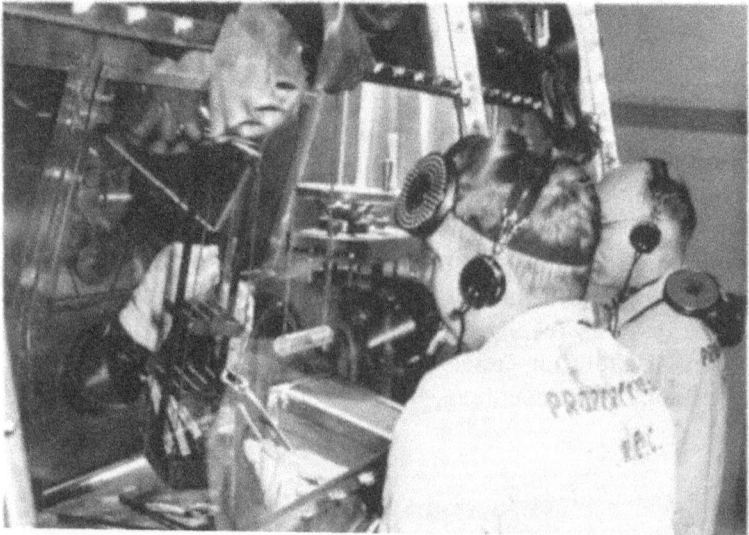

tron counters — are used routinely to reveal any plutonium contamination or other sources of hazardous radiation within the laboratory.

In a plutonium laboratory, workers usually wear protective clothing such as canvas or plastic booties and smocks or coveralls over their street clothes. The protective clothing is collected periodically and cleaned in a laundry with special facilities. Respirators are also worn whenever there is a possibility that plutonium might be inhaled, and special supplied-air masks are used if the contamination level is high.

A health monitor using a "PeeWee" (proportional alpha counter) to search for plutonium contamination during a routine survey of laboratory facilities. As little as a billionth of a gram of plutonium can be detected with this meter.

Safety Record   In spite of plutonium's unusually dangerous properties, people routinely work safely with it by paying careful attention to safety precautions. The experience record in atomic energy plants and laboratories shows fewer accidents than in hospitals, hotels, department stores, chemical plants, shipyards, auto plants and, of course, homes.

PERIODIC CHART OF THE ELEMENTS

1 H																	2 He
3 Li	4 Be											5 B	6 C	7 N	8 O	9 F	10 Ne
11 Na	12 Mg											13 Al	14 Si	15 P	16 S	17 Cl	18 Ar
19 K	20 Ca	21 Sc	22 Ti	23 V	24 Cr	25 Mn	26 Fe	27 Co	28 Ni	29 Cu	30 Zn	31 Ga	32 Ge	33 As	34 Se	35 Br	36 Kr
37 Rb	38 Sr	39 Y	40 Zr	41 Nb	42 Mo	43 Tc	44 Ru	45 Rh	46 Pd	47 Ag	48 Cd	49 In	50 Sn	51 Sb	52 Te	53 I	54 Xe
55 Cs	56 Ba	57-71 La* Series	72 Hf	73 Ta	74 W	75 Re	76 Os	77 Ir	78 Pt	79 Au	80 Hg	81 Tl	82 Pb	83 Bi	84 Po	85 At	86 Rn
87 Fr	88 Ra	89-103 Act† Series	(104)	(105)	(106)	(107)	(108)										

*Lanthanide Series

57 La	58 Ce	59 Pr	60 Nd	61 Pm	62 Sm	63 Eu	64 Gd	65 Tb	66 Dy	67 Ho	68 Er	69 Tm	70 Yb	71 Lu

†Actinide Series

89 Ac	90 Th	91 Pa	92 U	93 Np	94 Pu	95 Am	96 Cm	97 Bk	98 Cf	99 Es	100 Fm	101 Md	102 (102)	(103) Lw

*The transuranium elements (in shaded squares) are part of the actinide series of elements which as a group oc-
cupy a single square, at actinium (number 89) in the main figure. Plutonium, element 94, is in this series. The rare
earth (lanthanide) series of elements, also shown in a horizontal row, also occupies a single square (at lanthanum,
element 57) on the main chart.

ALPHABETICAL LIST OF ELEMENTS AND SYMBOLS

Element	Symbol	Element	Symbol	Element	Symbol	Element	Symbol
Actinium	Ac	Erbium	Er	Mercury	Hg	Samarium	Sm
Aluminum	Al	Europium	Eu	Molybdenum	Mo	Scandium	Sc
Americium	Am	Fermium	Fm	Neodymium	Nd	Selenium	Se
Antimony	Sb	Fluorine	F	Neon	Ne	Silicon	Si
Argon	Ar	Francium	Fr	Neptunium	Np	Silver	Ag
Arsenic	As	Gadolinium	Gd	Nickel	Ni	Sodium	Na
Astatine	At	Gallium	Ga	Niobium	Nb	Strontium	Sr
Barium	Ba	Germanium	Ge	(Columbium)		Sulfur	S
Berkelium	Bk	Gold	Au	Nitrogen	N	Tantalum	Ta
Beryllium	Be	Hafnium	Hf	Osmium	Os	Technetium	Tc
Bismuth	Bi	Helium	He	Oxygen	O	Tellurium	Te
Boron	B	Holmium	Ho	Palladium	Pd	Terbium	Tb
Bromine	Br	Hydrogen	H	Phosphorus	P	Thallium	Tl
Cadmium	Cd	Indium	In	Platinum	Pt	Thorium	Th
Calcium	Ca	Iodine	I	Plutonium	Pu	Thulium	Tm
Californium	Cf	Iridium	Ir	Polonium	Po	Tin	Sn
Carbon	C	Iron	Fe	Potassium	K	Titanium	Ti
Cerium	Ce	Krypton	Kr	Praseodymium	Pr	Tungsten	W
Cesium	Cs	Lanthanum	La	Promethium	Pm	(Wolfram)	
Chlorine	Cl	Lawrencium	Lw	Protactinium	Pa	Uranium	U
Chromium	Cr	Lead	Pb	Radium	Ra	Vanadium	V
Cobalt	Co	Lithium	Li	Radon	Rn	Xenon	Xe
Copper	Cu	Lutetium	Lu	Rhenium	Re	Ytterbium	Yb
Curium	Cm	Magnesium	Mg	Rhodium	Rh	Yttrium	Y
Dysprosium	Dy	Manganese	Mn	Rubidium	Rb	Zinc	Zn
Einsteinium	Es	Mendelevium	Md	Ruthenium	Ru	Zirconium	Zr

GLOSSARY

ADSORPTION   The process by which gas or liquid molecules become attached in thin layers to surfaces.

AIR SAMPLER   A device used to determine the amount of plutonium contamination in air. In some samplers, air is drawn continuously through a filter paper for a given length of time, such as a day. The filter paper is then analyzed to determine how much and what kind of radioactive material it collected. Other types of samplers give an immediate audible or visual warning whenever the amount of radioactive contamination in the air becomes dangerously high.

ALLOTROPE   Certain metallic elements can exist in more than one crystal structure form in different temperature and pressure ranges. These different forms are called allotropes.

ALLOY   A metallic or metallic-like substance composed of two or more elements, at least one of which is a metal.

ALPHA PARTICLE (RAY)   A type of nuclear radiation consisting of two protons and two neutrons, essentially the nucleus of a helium atom. It has a positive electrical charge that is twice as great as the negative charge of one electron. Alpha particles are thrown off at high velocity by many radioactive substances but have little penetrating power.

ANODE   In an electrical system, the pole or terminal to which electrons travel. In an electrolytic cell the anode is conventionally called the positive terminal; in a primary cell or storage battery it is called the negative terminal. (See also Cathode.)

ATOMIC NUMBER   The number assigned to each element on the basis of the number of protons found in the element's nucleus. Hydrogen has atomic number 1, helium atomic number 2, lithium, 3, etc.

ATOMIC WEIGHT (ATOMIC MASS)   Approximately the sum of the numbers of protons and neutrons found in the nucleus of an atom. Ordinary oxygen, for example, consists of a mixture of three isotopes having the atomic masses of 16, 17, and 18, but the isotope oxygen-16 is present in the mixture in such a large proportion, more than 99%, that the atomic weight of oxygen is given as 16. That particular isotope has 8 protons and 8 neutrons in its nucleus.

BETA–GAMMA SURVEY METER   An instrument used to detect the presence and determine the intensity of beta and gamma radiation.

BETA PARTICLE (RAY) A type of nuclear radiation which is essentially the same as an electron. It has a single electric charge and is thrown off at high velocity from the nuclei of some radioactive elements. Beta particles are more penetrating than alpha particles but less penetrating than gamma rays or X rays.

BORIDE A compound containing boron as the less positive element; for example, FeB and Fe_2B are two iron borides.

CARBIDE A compound containing carbon as the less positive element; for example, Ta_2C is a tantalum carbide.

CATHODE In an electrical system, the pole or terminal from which electrons travel. (See also Anode.)

CERAMIC A nonmetallic compound or mixture of compounds having a high melting point.

CHEMICAL ENERGY Energy produced by reactions between atoms; carbon and oxygen atoms, for example, combine to form CO_2 and produce chemical energy in the form of heat. (See also Nuclear Energy.)

COMPOUND A substance, usually of definite composition, formed by the chemical combination of two or more elements.

CONCENTRATE The valuable but impure mineral product obtained in the treatment of ore; the waste product is "tailing," which is discarded. Usually the concentrate is not the final product but must be treated further, for example to produce a pure metal.

CONDUCTIVITY The property of a material which describes the material's ability to transmit heat (thermal conductivity) or electricity (electrical conductivity). Electrical conductivity is the reciprocal of electrical resistivity. (See also Resistivity.)

CONTAMINATION Radioactive material, such as plutonium or plutonium oxide dust, which has been deposited in some place where it is not wanted, and particularly where its presence may be harmful.

COOLING PERIOD The period of time during which irradiated material loses its initial intense radioactivity. For example, the 2 to 4 months during which the irradiated uranium slugs from a plutonium-producing reactor are stored under water while their intense radioactivity becomes weaker. The water serves as a shield to confine the radiation and also cools the slugs.

COSMIC RAYS Radiations of many sorts which bombard the earth from outer space.

CRITICAL MASS   The smallest amount of fissionable material which, subject to the existing physical conditions, can undergo a nuclear chain reaction.

CRUCIBLE   A container for molten metal.

CRYSTAL STRUCTURE   Almost all elements and compounds solidify in orderly, repetitive, three-dimensional arrangements of atoms called crystals. The positions of the atoms and the distances between them in these crystals define the crystal structure, which is unique for each substance and for each allotrope of a substance.

CYCLOTRON   A device for speeding up charged atomic particles, such as protons and deuterons, by subjecting them to the combined action of a constant magnetic field and an alternating electrostatic field. The particles are whirled in a spiral path between the poles of a huge magnet, which greatly increases their velocities and thus increases their energies. These highly energetic particles are then directed so that they bombard the cyclotron target, usually to bring about nuclear changes in the target material.

CYCLOTRON TARGET   The material being bombarded by the beam of high-energy particles from a cyclotron. (See also Cyclotron.)

DAUGHTER ELEMENT   An element produced through the radioactive decay of a different (parent) element.

DECAY (RADIOACTIVE)   The gradual change of one radioactive element into a different element by the spontaneous emission of alpha, beta, or gamma rays. The end product, obtained usually after several stages of decay, is a nonradioactive, stable element.

DENSITY   That property of a substance which is expressed by the ratio of its mass to its volume, usually as grams per cubic centimeter (g/cm^3) or pounds per cubic inch ($lb/in.^3$).

DIODE   A vacuum tube consisting of only two electrodes, an anode and a cathode.

DISTILLATION   The process of heating a mixture of substances to evaporate the more volatile constituents which may then be collected and condensed to obtain a purer substance.

ELECTROLYTE   A solution that can conduct an electric current.

ELECTROLYTIC CELL   The assembly, including the container, electrodes, and electrolyte, which is used in the decomposition of compounds or the purification of materials by electrolysis.

ELEMENT   A pure substance consisting only of atoms having the same numbers of protons in their nuclei (the same atomic number).

FERTILE   A substance which is not fissionable itself but which can be converted into a fissionable material is fertile. For example, thorium-232 is not fissionable but can be converted into a fissionable material.

FISSILE   A substance that is fissionable, such as plutonium-239 or uranium-235, is fissile.

FISSION FRAGMENTS   The primary or initial elements formed as the result of a fission reaction.

FISSION PRODUCTS   The elements formed as the result of the radioactive decay of fission fragments, as well as the fission fragments themselves.

FISSIONABLE   Pertaining to the property of an atomic nucleus which causes it to split into two different nuclei, with an accompanying release of radioactivity and heat, when it absorbs a neutron.

GAMMA RAY   A highly penetrating type of nuclear radiation similar to X radiation, except that it comes from within the nucleus of an atom and, in general, has a shorter wavelength. (See also X ray.)

HALF-LIFE   The time required for half the atoms in a radioactive substance to disintegrate. The half-life is a characteristic property of each radioactive element and, depending on the element being considered, may vary from a millionth of a second to billions of years.

HEAT OF FUSION   The amount of heat needed to change 1 gram molecular weight of a substance from the solid to the liquid state, without increasing the temperature of the substance.

HEAT OF TRANSFORMATION   The amount of heat needed to cause 1 gram molecular weight of an allotrope of a substance to transform to the allotrope which is stable in the next higher temperature range, without increasing the temperature of the substance.

HEAVY ELEMENTS   Refers to the relative weights of the atoms. Hydrogen is the lightest and uranium the heaviest of the naturally occurring elements. Hence, those elements near uranium in the Periodic Table are heavy elements, and those near hydrogen are light elements.

HERMETICALLY SEALED   Completely sealed, airtight.

HYPOTHESIS   A preliminary and usually incomplete explanation of some natural principle.

INERT   Nonreactive, does not take part in chemical reactions.

INHALE   To breathe into the lungs.

INTERMETALLIC COMPOUND   A compound consisting of two or more metallic elements. (See also Compound.)

ION EXCHANGE   A chemical process involving the reversible interchange of ions between a solution and a solid substance (ion-exchange resin).

IRRADIATED   Exposure to some form of radiation; for example, a uranium fuel element that has been exposed to neutron activity in a nuclear reactor.

ISOTOPE   Elements may consist of a number of isotopes. Isotopes of a given element have the same atomic number (same number of protons in their nuclei) but different atomic weights (different numbers of neutrons in their nuclei). Uranium-238 and uranium-235 are isotopes of uranium. The uranium-238 atom has 92 protons and 146 neutrons in its nucleus; the uranium-235 nucleus contains 92 protons and 143 neutrons.

LATTICE ARRANGEMENT OR PILE   The geometrical arrangement of the solid fissionable and nonfissionable material in a nuclear reactor; for example, the arrangement of the uranium and graphite blocks in the first chain-reacting assembly, the Chicago chain-reacting pile.

LIGHT ELEMENTS   (See Heavy Elements.)

MASS   A measure of the amount of matter in a body, expressed in grams, pounds, etc.

METALLURGY   The science and technology of metals.

MICROHM   One millionth of an ohm, a unit of electrical resistance.

MOLTEN-METAL EXTRACTION   The separation of one metal from a mixture by the dissolving action of a different molten metal.

MULTIPLICATION FACTOR   The ratio of the number of neutrons produced in any one generation to the number that existed in the immediately preceding generation.

NEUTRON   One of the basic particles which make up an atom. A neutron and a proton have about the same weight, but the neutron has no electrical charge. Slow neutrons have relatively low energies; fast neutrons have high energies.

NEUTRON COUNTER   An instrument used to detect the presence and determine the intensity of neutron radiation.

NEUTRON FLUX   The quantity of neutron energy being emitted per unit of time and across a unit of surface area.

NEUTRON SOURCE   A quantity of neutron-emitting radioactive substance packaged conveniently for scientific or industrial use.

NITRIDE   A compound containing nitrogen as the less positive element; for example, UN is uranium nitride.

NUCLEAR ENERGY   Energy produced by a nuclear reaction (fission or fusion) or by radioactive decay. (See also Chemical Energy.)

NUCLEUS   The core of the atom, where most of its mass and all its positive charge is concentrated. It consists of protons and neutrons (except ordinary hydrogen).

ORDER OF MAGNITUDE   A numerical comparison expressed as a multiple of some unit taken as standard; for example, 10, 100, and 1000 are different orders of magnitude.

OXIDATION STATE   The condition of an atom or ion which determines its ability to take on or give up electrons. Oxidation involves the removal of electrons from an atom or ion.

OXIDE   A compound containing oxygen as the less positive element; for example, FeO, Fe_2O_3, and Fe_3O_4 are different types of iron oxides.

OXIDIZE   To unite with oxygen. (See also Oxidation State.)

PERIODIC TABLE   The arrangement of the elements according to their atomic numbers and in such a way as to reveal the periodic or regular variations in many of their properties.

PHASE   A single, distinct kind of substance existing in one state. Ice and water are the solid and liquid phases of the compound H_2O.

PICKLING   The cleaning of a metal surface by chemical or electrochemical means.

PILE   A nuclear reactor named after the earliest types of reactors which were actually piles of graphite and uranium blocks.

PITCHBLENDE   A mineral containing uranium.

PRODUCTS   The substances that are produced in a chemical or nuclear reaction.

PROPORTIONAL ALPHA COUNTER   An instrument used to detect the presence and determine the intensity of alpha radiation.

PROTON   One of the basic particles which make up an atom. The proton has a positive electrical charge equivalent to the negative charge of an electron and a mass similar to that of a neutron.

RADIOACTIVE   Giving off energy in the form of alpha, beta, or gamma rays through the disintegration of atomic nuclei.

RARE GASES   Gaseous elements, such as krypton and xenon, which are normally present in extremely small proportions, about 1 part or less per million, in the atmosphere.

REACTANTS   The substances that enter into a chemical or nuclear reaction.

REACTIVE   The tendency for one substance to combine chemically with another.

REACTIVITY   A measure of the departure of a nuclear reactor from criticality. It is equal to the multiplication factor minus one and is thus zero precisely at criticality. If there is excess reactivity (positive reactivity), the reactor power will rise. Negative reactivity will result in a decreasing power level.

REACTOR   An assembly of fissionable and nonfissionable materials for producing nuclear energy or nuclear materials.

REACTOR CORE   The central portion of the reactor, which includes the fissionable material.

REDUCING AGENT, REDUCTANT   A substance that deoxidizes (or removes the oxygen from another substance) and thus becomes oxidized itself or that acts to produce a metallic element from its compounds.

REFRACTORY   Difficult to melt, high-melting-point materials. Tantalum and tungsten, for example, are two refractory metals.

RESISTIVITY   The resistance to the flow of electricity in a rod of unit length and unit cross-sectional area. (See also Conductivity.)

RESPIRATOR   A protective device covering the mouth and nose. It has filters which trap poisonous particles or vapors in the atmosphere.

SALT   A chemical compound containing a metallic and a nonmetallic element, formed by the chemical reaction between an acid and a substance that replaces some or all the hydrogen of the acid.

SALT EXTRACTION   The separation of one element from a mixture by the dissolving action of a molten salt.

SELECTIVE PRECIPITATION   A chemical process in which dissolved substances can be separated from each other by adjustment of the conditions of the solution so that one of the substances becomes insoluble and settles out of the solution.

SHIELDING   A protective wall or barrier, usually of a dense solid or liquid material, which prevents the spread of radiation from radioactive materials to the surrounding areas.

SILICIDE   A compound containing silicon as the less positive element; for example, $CaSi_2$ is calcium silicide.

SOLVENT EXTRACTION   A chemical process whereby a liquid solvent is used to dissolve one of the substances in a mixture and thus separate it from the mixture. In a subsequent step the substance is removed from the solvent.

SPECIFIC HEAT   The quantity of heat needed to raise the temperature of 1 gram of a given substance 1°C.

STABLE   Not readily changed chemically; nonradioactive.

SUPPLIED-AIR MASK   A protective device similar to a respirator but covering the entire face and having an independent source of pure air, such as a compressed air tank, for breathing.

THERMIONIC   Related to thermions, which are electrically charged particles emitted by an incandescent substance.

THERMOELECTRICITY   Electricity produced by the direct action of heat on thermocouples.

TNT   Trinitrotoluene, a chemical explosive.

TRACER   A small amount of a radioactive isotope used in following an element through various chemical and physical changes. Usually, the amount of the radioactive element involved is so small as to be invisible and unweighable.

TRANSFORMATION   The changing of one allotrope of an element into a different allotrope of the same element.

ULTRAMICROSCALE   Pertains to extremely small-scale operations.

VALENCE   The power one element has of combining chemically with another, based on the number of hydrogen atoms with which the element will combine or replace. Oxygen, which has a valence of 2, combines with hydrogen in the ratio of 1 atom of oxygen to 2 atoms of hydrogen (H_2O).

VAPOR PRESSURE   The pressure exerted by the confined vapor in equilibrium with its solid or liquid form.

X RAY   Highly penetrating radiation similar to gamma rays but usually of longer wavelength and originating outside the nucleus of the atom. (See also Gamma Ray.)

SUGGESTED REFERENCES

Books

The Metal Plutonium, A. S. Coffinberry and W. N. Miner (Eds.), The University of Chicago Press, Chicago, Ill., 1961, 446 pp., $9.50. Based on the First International Conference on Plutonium, this book contains information on the history, physical properties, and proposed reactor uses of plutonium.

The Rare Metals Handbook, 2nd ed., C. A. Hampel (Ed.), Reinhold Publishing Corporation, New York, 1961, 715 pp., $22.50. Chapter 18, consisting of 57 pp. authored by W. N. Miner, A. S. Coffinberry, F. W. Schonfeld, J. T. Waber, R. N. R. Mulford, and R. E. Tate, summarizes physical property information about plutonium. Short sections on sources, production, extraction, metal preparation, chemical properties, toxicity, metallography, fabrication, and applications are also included.

The Transuranium Elements, Glenn T. Seaborg, Addison-Wesley Publishing Company, Inc., Reading, Mass., 1958, 328 pp., $7.00. This volume contains much historical information about plutonium as well as information about its chemical and nuclear properties.

Atomic Energy for Military Purposes, H. D. Smyth, Princeton University Press, Princeton, N. J., 1945, 264 pp., $4.00. The development of the atomic bomb between 1940 and 1945 is described in detail. Some historical and nuclear property information on plutonium is included.

Proceedings of the International Conference on the Peaceful Uses of Atomic Energy, United Nations, New York, 1956. Volumes 7 and 9 contain plutonium information.

Proceedings of the Second International Conference on the Peaceful Uses of Atomic Energy, United Nations, Geneva, 1958. Volumes 5, 6, 7, 13, 14, 17, and 23 contain plutonium information.

Encyclopaedia Britannica, article on plutonium, pp. 92A−92D, Volume 18, William Benton, Publisher, Chicago, Ill., 1962.

McGraw-Hill Encyclopedia of Science and Technology, article on plutonium, pp. 422-25, Volume 10, McGraw-Hill Book Company, Inc., New York, 1960.

Articles

Transuranium Elements, Glenn T. Seaborg, *Science,* 104: 379-86 (Oct. 25, 1946).

Dr. Seaborg and His Man-Made Atoms, W. S. Barton and M. M. Hunt, *Science Illustrated,* 4: 46-8+ (February 1949).

The Atomic Bomb, *Life*, 28(9): 90-7 (Feb. 27, 1950).

Plutonium Saga, *Scientific American,* 200: 66+ (February 1959).

Plutonium for Reactors, *Scientific American,* 201: 65-6 (July 1959).

Production of Plutonium Metal, Elspeth W. Mainland with D. A. Orth, E. L. Field, and J. H. Radler, *Industrial and Engineering Chemistry,* 53: 685-94 (September 1961).

Nucleonics, 21: 37-52 (January 1963), contains a series of articles on plutonium as a nuclear fuel.

Plutonium Foundry Practices, J. W. Anderson and W. J. Maraman, *American Foundrymen's Society. Transactions,* 70: 1057-72 (1963).

Reports

Plutonium as a Power Reactor Fuel, Proceedings of the American Nuclear Society Topical Meeting held at Richland, Washington, September 13 and 14, 1962. AEC Report HW-75007, Hanford Atomic Products Operation, Richland, Washington (December 1962) 542 pp. Available from the Clearinghouse for Federal Scientific and Technical Information, 5825 Port Royal Road, Springfield, Virginia 22151. $6.00.

Motion Pictures

Available for loan without charge from the AEC Headquarters Film Library, Division of Public Information, U. S. Atomic Energy Commission, Washington, D. C. 20545 and from other AEC film libraries.

Plutonium Fuel Fabrication for MTR, 11 minutes, color and sound, 1958. Produced by the Hanford Atomic Products Operation, General Electric Company. The Materials Testing Reactor (MTR) at AEC's National Reactor Testing Station, Idaho, has been operated utilizing plutonium as the entire fissionable fuel charge. This technical film details the fabrication of this charge in the plutonium metallurgy laboratories of AEC's Hanford Works, Richland, Washington.

Plutonium Metal Preparation, 12 minutes, black and white, sound, 1958. Produced by AEC's Los Alamos Scientific Laboratory. This technical film shows the process and equipment used in converting plutonium from a nitrate solution to elemental metal.

Fabrication of Plutonium Disks, 13 minutes, black and white, sound, 1958. Produced by AEC's Los Alamos Scientific Laboratory. This is a companion film to *Plutonium Metal Preparation.* The film describes glove box work used at Los Alamos Scientific Laboratory in shaping toxic material for criticality studies in reactor development.

The Fuel of the Future, 29 minutes, sound, black and white, 1965. Produced by Ross-McElroy Productions for the National Educational Television and Radio Center under a grant from Argonne National Laboratory. Special precautions and techniques employed in work with plutonium are shown in a unique engineering laboratory, the Argonne Fuel Fabrication Facility, where work is performed within sealed glove boxes under an inert atmosphere. The manufacture of experimental reactor fuel pins containing plutonium is illustrated step-by-step.

PHOTO CREDITS

Cover courtesy J. R. Morgan, Los Alamos Scientific Laboratory (LASL)

Page

7 *Chemical and Engineering News* (Seaborg and McMillan)
   Washington University (Kennedy and Wahl)
10 General Electric Corporation
17 LASL
28 LASL
29 LASL
30 Argonne National Laboratory, Idaho Division
31 Oak Ridge National Laboratory (top)
   Monsanto Research Corporation
32 LASL
36 LASL
38 LASL
39 LASL